Pictorial Zoology

№ 6 Plate 1

Passenger Pigeon

Ectopistes migratorius

LASZLO LAYTON

Pictorial Zoology

NAZRAELI PRESS
in association with
Peter Fetterman Gallery

Of Loss and Memory

The Artistry of Laszlo Layton

Laszlo Layton has devoted his artistic passion to the preservation and memory of the animals and specimens he truly loves, many of which have sadly disappeared forever. He himself is an endangered species in the contemporary photographic arena so dominated by the tricks of the digital world. He really belongs in another century in his employment of labor intensive, hand-crafted artistry to create his sublime images of nature.

Yes, he does drive a car and wear jeans—but my image of him, before we actually met, was of someone riding in a horse and buggy, wearing a velvet topcoat and stiff-winged collar, leaving the British Museum after having spent a day there researching to join Charles Darwin in his club to discuss current scientific matters, and sipping a glass of port. Darwin would then ask Laszlo to accompany him on his voyage on the HMS *Beagle*, which he would excitedly agree to do. He would also join John James Audubon on his travels, the latter carrying his sketch book, paint and canvasses, and a gun and Laszlo heroically lifting his heavy view-camera on his back.

He belongs in an Arts and Crafts house with William Morris as his neighbor, and no doubt would have been exhibited by Alfred Steiglitz at Gallery 291, and been published by him in *Camerawork*. His photographic predecessors, Gertrude Kasebier, Clarence White, George Seeley, F. Holland Day, and Edward Steichen would have welcomed him as a fellow artist.

But we are fortunate that that he is with us now, to both remind us what we have lost, and what we may still preserve. Who cannot be charmed and moved by the spirit of his *Little Egret*, *Blue Morpho Butterfly*, *Scarlet Ibis*, and the fragility of his *Emu Egg*?

He has tamed the ferocity of his *Snow Leopard*, his *Jaguar*, and *Giant Panda* to show their own vulnerability at the often cruel hands of man.

Embrace this exquisite book and spread his subtle message for all to be vigilant for the beauty that still surrounds us.

— *Peter Fetterman* —
Peter Fetterman Gallery

№ 6
Pl. 85
Lion
Panthera leo

N°4
Pl. 58
Vervet Monkey
Chlorocebus
pygerythrus

A.

Pl.84

Carmine Bee-eater

Merops nubicus

A.
Pl.88
King Rail
Rallus elegans

A.
Pl. 64
Violet Spider Conch
Lambis violacea

A.
Pl. 9
Snowy Owl
Nyctea scandiaca

N° 6

Pl. 12

Guianan Cock-of-the-Rock

Rupicola rupicola

Emu (egg)

Dromiceius novaehollandiae

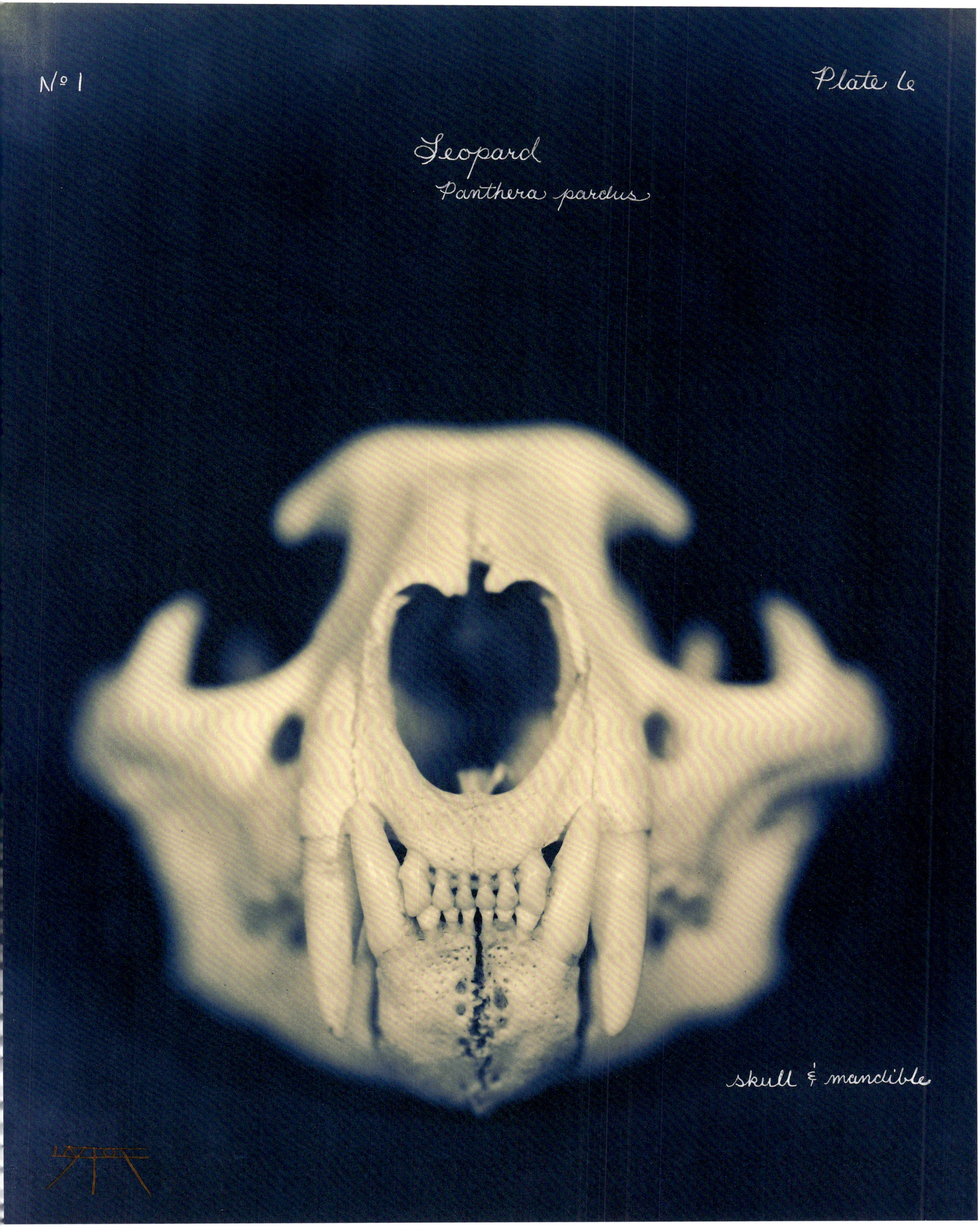

No 1
Plate 6
Leopard
Panthera pardus
skull & mandible

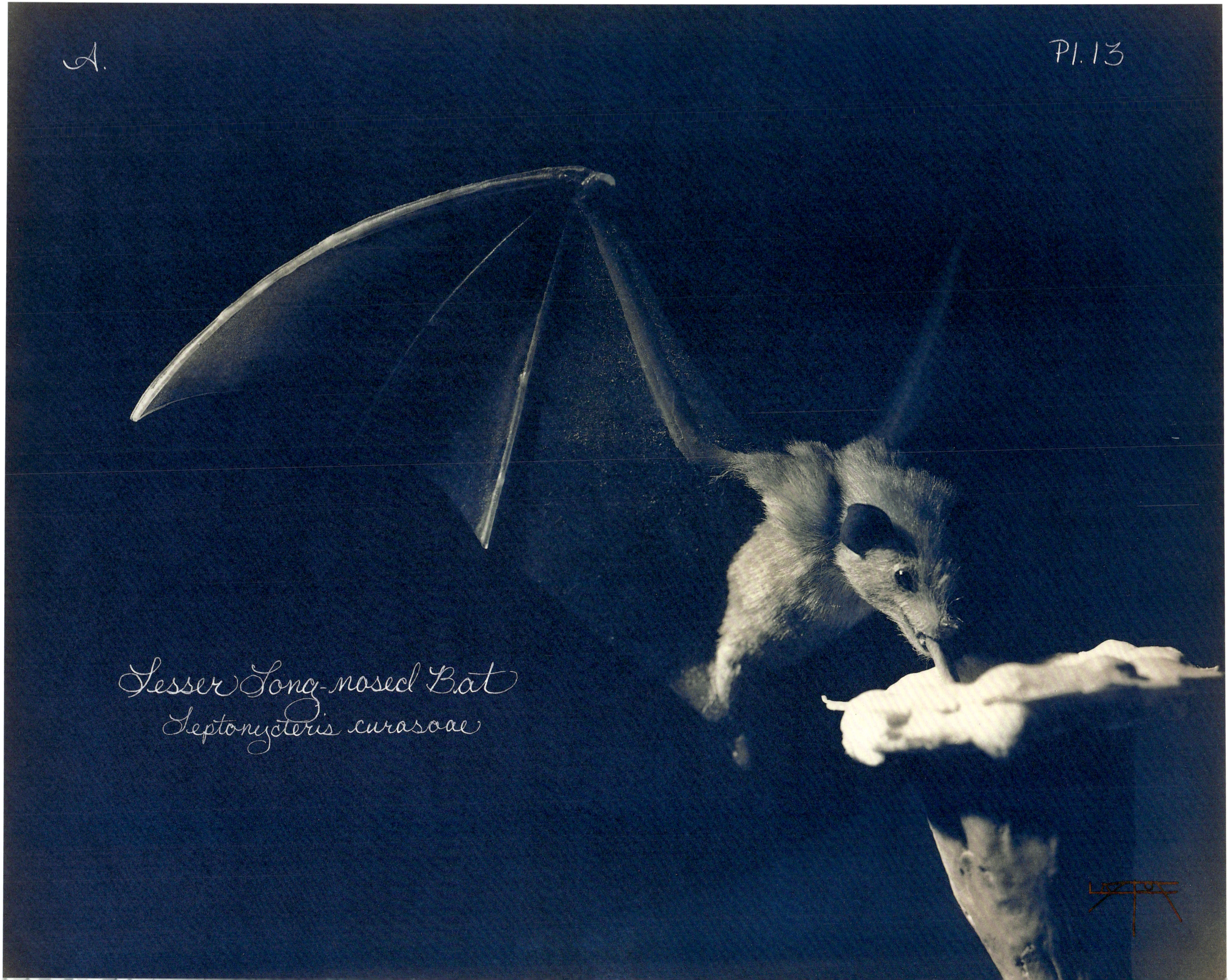
A.
Pl. 13
Lesser Long-nosed Bat
Leptonycteris curasoae

No 10
Pl. 18
Great Spotted Kiwi
Apteryx haastii

No 6

PL. 17

Saffron Murex
Chicoreus crocatus

№ 1
Pl. 89
Soemmerring's Gazelle
Gazella soemmerringii

A.
Pl. 33
Rock Ptarmigan
Lagopus mutus

B. American Thorny Oyster *Spondylus americanus* Pl. 14

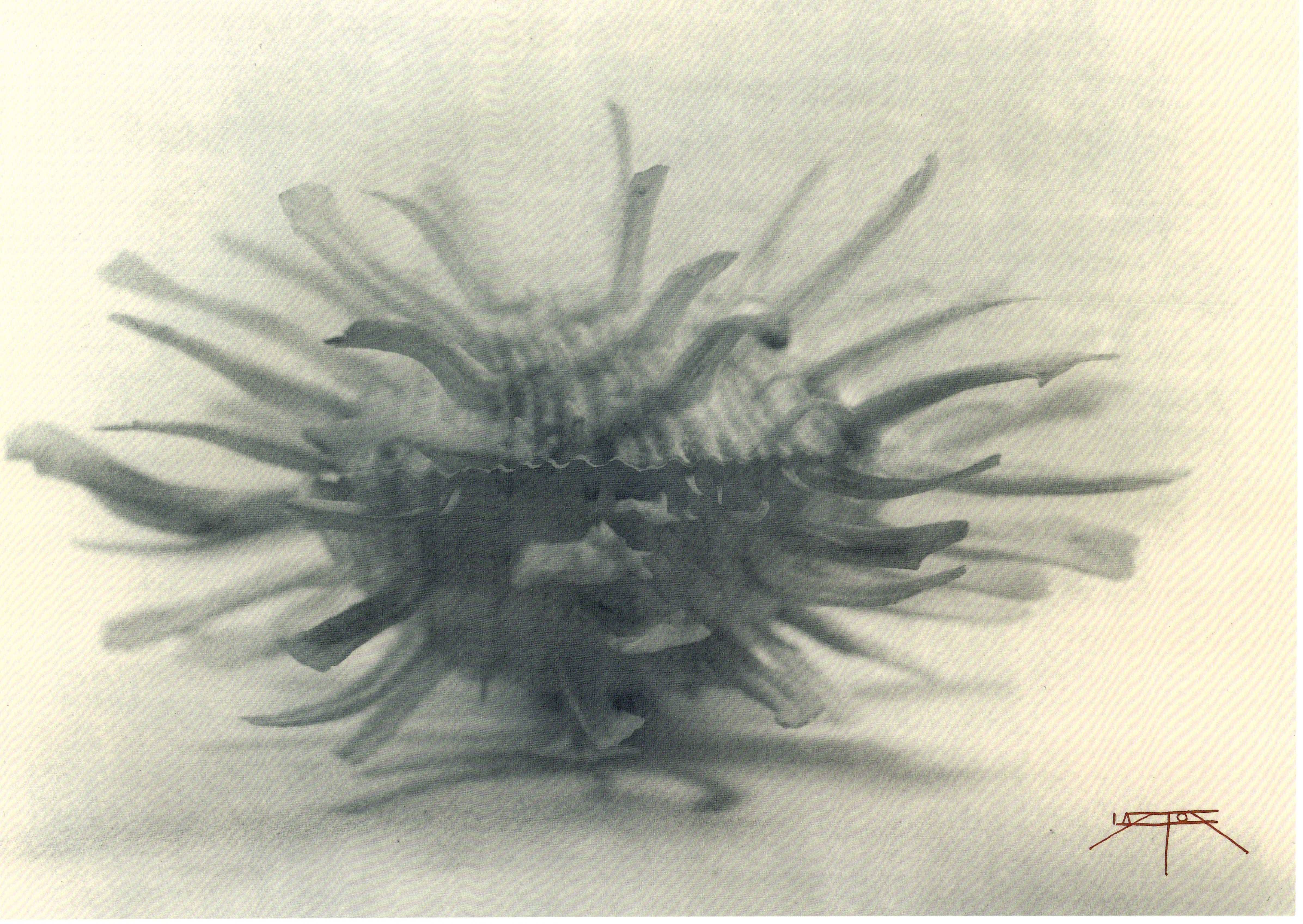

No 6
Pl. 48
Little Egret
Egretta garzetta

No 3

Pl. 35

Slate Pencil Urchin

Heterocentrotus mammillatus

№ 1
Pl. 100
Carolina Paraquet
Psittacus carolinensis

No 3

Pl. 46

№ 7

Pl. 26

Douroucouli
Aotus trivirgatus

No 4

Pl. 75

Ivory-billed Woodpecker
Campephilus principalis

No6
Pl.42
North American Porcupine
Erethizon dorsatum
(albino)

№4
Pl.41
Raccoon
Procyon lotor

№ 6
PL.20
Black Grouse
Tetrao tetrix

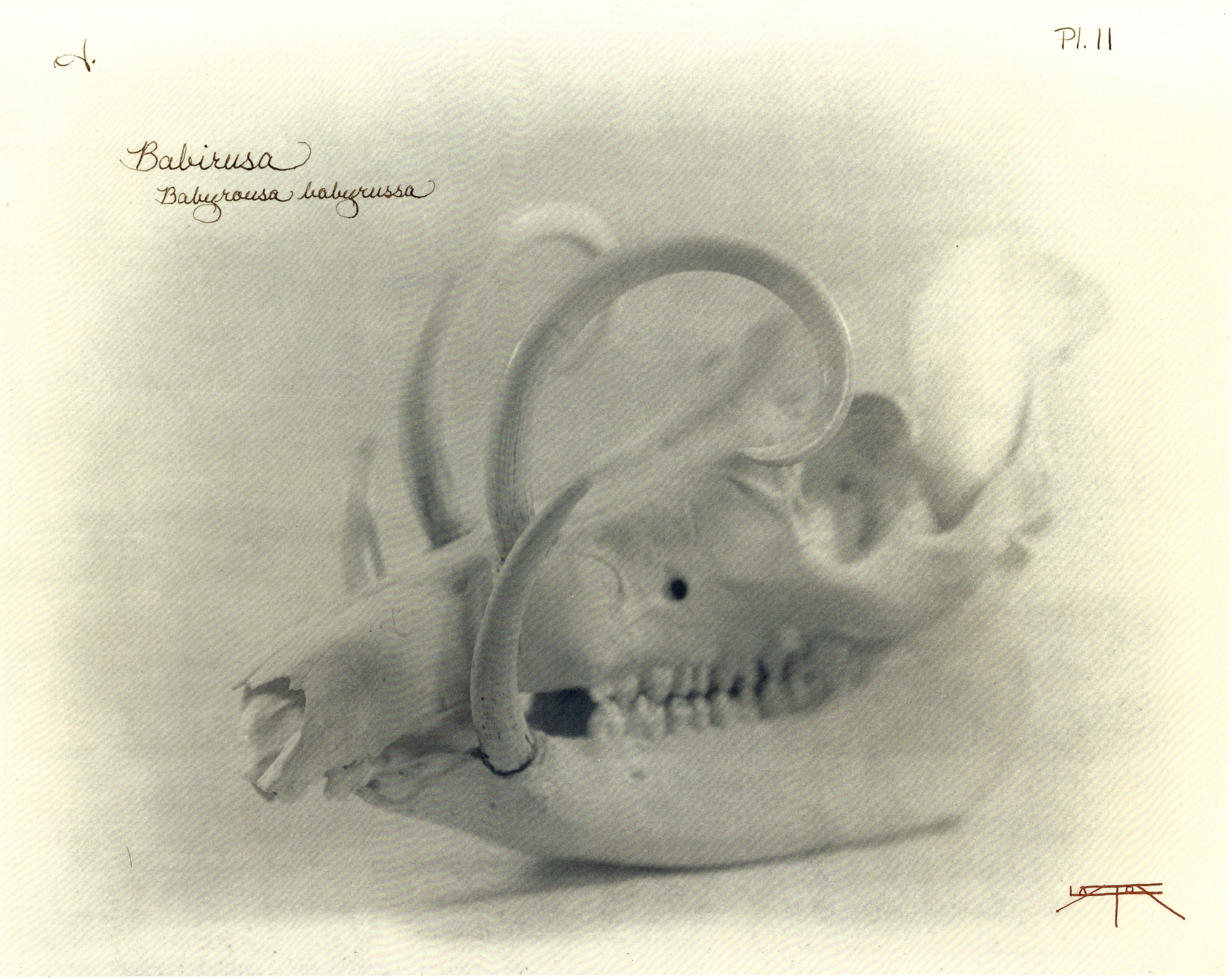

Pl. 11
Babirusa
Babyrousa babyrussa

No 4
Plate 3
Blue Morpho Butterfly
Morpho nestira

№4
Pl.60
Grivet Monkey
Chlorocebus aethiops

No 6
Pl. 50
Giant Panda
Ailuropoda
melanoleuca

No 6

PL. 31

Great-tailed Grackle

Quiscalus mexicanus

No 1

Pl. 105

Oriental Pied Hornbill

Coronatus malabaricus

No 4
PL. 24
Spotted Hyaena
Crocuta crocuta

A.
Pl. 102
Tibetan Takin
Budorcas taxicolor tibetana

A.
Pl.80
Black-tailed Prairie Dog
Cynomys ludovicianus

No 7
PL. 16
Kakapo
Strigops habroptilus

No 8

Pl. 21

Regal Thorny Oyster

Spondylus regius

Nº 5
Pl. 72
Serval
Leptailurus serval

A.
Pl. 66
Smew
Mergellus albellus

A.
Pl.86
Roseate Spoonbill
Ajaia ajaja

Bat-eared Fox
Otocyan megalotis

№ 6
PL. 30
Jaguar
Panthera onca arizonensis

№3
Pl. 83
Nine-banded Armadillo
Dasypus novemcinctus

B.

Pl. 53

No 2

Pl. 43

Quetzal

Pharomachrus mocino

Porcupine Fish
Cyclikthys orbicularis

A.

Pl. 96

Little Blue Heron
Egretta caerulea

B.
Pl. 51
Argonaut
Argonauta argo

B.
Pl. 103
Chilean Flamingo
Phoenicopterus Chilensis

No 6

Pl. 15

Cacomistle

Bassariscus astutus

B.
Pl. 61
River Otter
Lontra canadensis

No 3
Pl. 56
Zebra Duiker
Cephalophus zebra

№ 6

Pl. 79

Crowned Pigeon

Goura cristata

No 8
Pl. 22
Lesser Prairie Chicken
Tympanuchus pallidicinctus

No 2
Pl.77
Saiga
Saiga tatarica

PL. 5
A.
Scarlet Ibis
Eudocimus ruber

A.
Pl. 91
Dorcas Gazelle
Gazella dorcas

B. Pl. 93

B.
Pl.90
Emerald Toucanet
Aulacorhynchus prasinus

The breadth of biodiversity represented by this project would not have been possible without the kind invitation and generous cooperation of the following museums and their staff: Richard White, International Wildlife Museum, Tucson, Arizona; Dr. Angel Valdes, Dr. James H. McLean, Lindsey T. Groves, Malacology Department, Dr. Joel W. Martin, George E. Davis, Dr. Gordon Hendler, Crustacea & Echinoderm Departments, Natural History Museum of Los Angeles County, Los Angeles, California; Dr. Linnea S. Hall, Rene Corado, Western Foundation of Vertebrate Zoology, Camarillo, California. My sincerest gratitude for allowing me such unrestricted access to your collections.

With thanks and appreciation to Terry Etherton, Joseph Bellows, Rixon Reed and Sam Lee for all your efforts on my behalf.

Special thanks to Peter Fetterman and Chris Pichler. Peter—I am forever grateful for your guidance, support of my work, belief in me as an artist, and genuine friendship. Chris—your hospitality and artistry made working on the design of this book one of the great experiences of my life, and I could not be more proud of the beautiful result.

Love and Aloha to my parents, Joyce Maldonado and Ernie Menehune, my family and my dear friends.

To my partner, Cambrian James, for your encouragement and unconditional love—this book is for you.

—*Laszlo Layton*—

Gold Canyon, June 2012